The Moon

Earth's satellite and
nearest neighbour
in space

Diameter
3476 km
0.27 of Earth's mean diameter

Mass
0.012 of Earth's mass

Density
3.33 in terms of water
0.60 of Earth's mean density

MARE FRIGORIS

Lunokhod 1 ▼

MARE IMBRIUM

MARE
SERENITATIS

▼ Lunokhod 2

Apollo 15 ▼

▼ Apollo 17

▼ Luna 13

MARE
CRISIUM

MARE
VAPORUM

Luna 24 ▼

Copernicus

MARE
TRANQUILLITATIS

▼ Luna 9

▼ Luna 20

OCEANUS
PROCELLARUM

Surveyor 5 ▼ ▼ Ranger 8

▼ Luna 16

▼ Surveyor 6

▼ Apollo 11

MARE
FECUNDITATIS

Apollo 12 ▼
Surveyor 3 ▼

Surveyor 1 ▼

▼ Apollo 14

MARE
COGNITUM

▼ Ranger 7

▼ Apollo 16

▼ Ranger 9

MARE
NECTARIS

MARE
NUBIUM

MARE
HUMORUM

Velocity in orbit
3680 km per hour

Surface temperature
120°C Sun at zenith
−153°C at night

▼ Surveyor 7
Tycho

Librations (variable movements)
7°54′ lateral displacement in longitude
6°50′ vertical displacement in latitude

Surface visible from Earth
59.0% maximum
41.0% always visible during lunar month
18.0% sometimes visible due to librations

Distance from Earth
404 336 km maximum
354 340 km minimum
382 176 km mean

Period of rotation (lunar month)
29.530588 days, new Moon to
new Moon
27.321661 days, in relation to stars

2 The Moon's nearside, showing principal topographic features and landing sites of exploration spacecraft

The growth of knowledge

Since earliest times Man has pondered the face of the Moon—whether impelled by mystic awe, by romantic associations, or by a growing scientific curiosity. Centuries before the birth of Christ the ancient philosophers of China, Egypt, Babylon and Greece first roughly established some of the Moon's astronomical constants, especially those in relation to the Earth. Thales of Miletus (about 600 BC) correctly deduced that the Moon reflected light from the Sun, although the Moon's changing phases were more clearly attributed to the Sun's illumination by Pythagoras (580–500 BC) and Aristotle (384–322 BC). Democritus (460–370 BC) first suggested that the lunar surface markings were due to the presence of great mountains and valleys. Aristarchus (310–230 BC) established the relative dimensions of Earth and Moon, later modified by Eratosthenes (275–194 BC) after more-exact calculations of the Earth's circumference and radius. About AD 150 Ptolemy determined the approximate distance between Earth and Moon.

The work of Copernicus, Tycho and Kepler during the 16th and 17th centuries more closely defined the Moon's position and motions in a Sun-centred planetary system, but it was Galileo's account (1610) of the Moon's surface as first seen through a telescope which marked the beginning of systematic scientific study. By the end of the 17th century telescopic observations of the Moon had led to the production of more than twenty-five maps of its visible surface. Amongst the most outstanding were those drawn by Langrenus (1645), Rheita (1645), Helvelius (1647), Riccioli (1651) and Cassini (1680). It was Riccioli who named many of the major lunar features, like the craters Copernicus and Tycho, and Mare Tranquillitatis.

Observers of the 18th and 19th centuries continued to refine the standards of lunar mapping, but it was the later rapid development of photography and the improvements in astronomical telescopes which laid the foundations of our present-day knowledge. New techniques of observation and study developed during the first half of the 20th century included radio, infrared and radar-reflection studies, alpha and gamma-ray spectroscopy, with surface luminescence, polarisation and reflectivity optical methods.

Space-age exploration began with the spectacular advances in rocketry, computers and instrument technology which followed the 1939–45 World War. The Moon, because of its close proximity to Earth, naturally became the first target for remote-controlled and manned spacecraft investigations. During the nineteen-sixties American scientists expended great efforts to achieve a manned landing on the Moon 'by the end of the decade'. First, three crash-landed Ranger spacecraft transmitted television pictures of the Moon's surface during the fifteen minutes before impact. Then five soft-landed Surveyor spacecraft not only scanned the surrounding lunar terrain with television cameras, but also carried out a number of physical experiments, remotely controlled from Earth, to ascertain the nature of the Moon's surface. Five Orbiter spacecraft meanwhile completed a high-resolution photographic record of the Moon's surface, from orbital distances ranging between 50 and 6000 kilometres.

Having launched the modern era of Moon exploration in 1959, with fly-by and crash-landed Luna spacecraft, the Russians were first (in January 1966) with close-up photography of its surface from the soft-landed Luna 9. Later Russian exploration by orbiting and soft-landed Zond and Luna spacecraft, somewhat overshadowed during the same period by American achievements, culminated in another 'first' (in September 1970) with the automatic sampling and return to Earth by Luna 16 of about 100 grams of the Moon's surface soil. Further diversification of Russian effort was provided by the successful landing and operation, remotely controlled from Earth, of two mobile Lunokhod vehicles. By this time, however, twelve American Apollo astronauts had walked upon the Moon (pp 10–21) and had brought back to Earth over 386 kg of lunar rock samples.

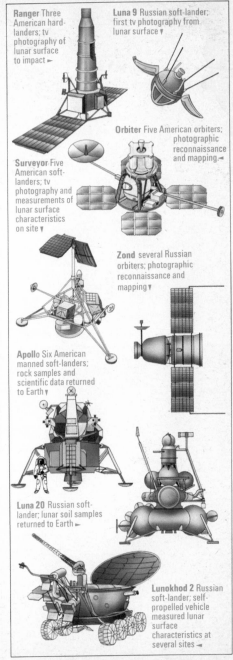

Ranger Three American hard-landers; tv photography of lunar surface to impact ►

Luna 9 Russian soft-lander; first tv photography from lunar surface ▼

Orbiter Five American orbiters; photographic reconnaissance and mapping ◄

Surveyor Five American soft-landers; tv photography and measurements of lunar surface characteristics on site ▼

Zond several Russian orbiters; photographic reconnaissance and mapping ▼

Apollo Six American manned soft-landers; rock samples and scientific data returned to Earth ▼

Luna 20 Russian soft-lander; lunar soil samples returned to Earth ►

Lunokhod 2 Russian soft-lander; self-propelled vehicle measured lunar surface characteristics at several sites ◄

3 The principal lunar exploration spacecraft

4 Mare Imbrium, a lava-filled impact basin 1000 km in diameter

5 Cratered terra south of Mare Nubium

The lunar landscape
Seen from Earth

The Moon revolves once on its axis each time it orbits the Earth, thus always presenting the same face to Earthbound observers. However, even to the unaided eye this unchanging face shows two contrasting types of landscape—dark, plain-like areas of low relief, and brighter, decidedly more rugged regions which cover about two-thirds of the surface. Early astronomers mistakenly referred to the smooth dark areas as *maria* (or 'seas'), giving the name *terrae* (or 'lands') to the bright upland regions. The terms have persisted since, even though the Moon's surface has long been known to be completely waterless.

Through binoculars or telescopes the terrae are seen to be mountainous, densely cratered regions. Craters of all sizes up to about 250 km across are scattered over the surface in great profusion, frequently overlapping one another. Some are well-defined circular features with rampart-like walls and lofty central mountains, others are scarcely discernible degraded structures, often modified or partly obliterated by super-imposed craters. Here and there mountains and valleys are intersected by scarp-like features and linear clefts or '*rilles*' (p 8). Mountain peaks in some highland regions rise 8000 m above the surrounding terrain. Mountain ranges form 'ringwall' borders to some maria, like those which extend in a great curve 600 km long, partly enclosing Mare Imbrium (fig 4).

The maria vary from almost circular features surrounded by mountains, like Imbrium and Crisium, to the more irregular-shaped regions like Tranquillitatis and Oceanus Procellarum. Their dark surfaces are deceptively smooth, broken only by a few large craters, but telescope observations reveal smaller craters and many irregularities, such as isolated peaks and hills, linear ridges and rilles. The maria are younger than the terrae whose cratered surfaces may be relics of a primordial landscape.

Two different processes have been responsible for the present-day landscape: crater-forming impact of countless meteorites and other objects from space on the Moon's atmosphere-free surface; and igneous and volcanic activity from within the Moon. Early in its history the Moon was subjected to intensive meteorite bombardment, including catastrophic basin-excavating impacts of massive asteroidal bodies. These huge basins were subsequently filled by tremendous surface outpourings of volcanic lava to form the maria. That they were not all formed at the same time can be deduced from their shapes and marginal relationships: the most ancient basins, including Nubium, Tranquillitatis and Procellarum, have lost most traces of original impact outlines whereas those excavated later, like Crisium and Imbrium, still retain circular shapes and bordering mountains. Changes in the lunar landscape since the last outpouring of mare lava have been mostly brought about by meteoritic and cometary impacts on a much reduced scale over a long period of time.

The farside

Man's first glimpse of the Moon's farside came with spacecraft pictures transmitted to Earth in 1959. Later, more detailed photographic mapping of the Moon by orbiting spacecraft revealed a mostly terra-type landscape on the farside, formed almost entirely by heavily cratered highlands (fig 6). Although lacking the large, lava-filled marial basins so characteristic of the nearside, the farside displays several large crater-like basins, some of which appear to have smooth, dark mare-type floors. The most striking of these is Mare Orientale (fig 7, lower right), a small part of which the Moon's libration movements occasionally render visible from Earth. It is a huge circular basin, almost 1000 km across, structurally outlined by several concentric mountain ranges rising to heights of 6000 m above base level. Many features of Orientale show it to be one of the youngest of the large impact basins formed early in the Moon's history, comparable in origin and size with Imbrium, but differing in relief pattern and less extensive lava infilling. Amongst other small mare basins of impact origin are Moscoviense, 400 km in diameter (fig 7, upper left), and Tsiolkovsky, a prominent crater 180 km in diameter, floored by dark lavas through which lofty central mountains rise in bright contrast (fig 33).

Why are the principal features of nearside and farside so different? Large impact basins are evenly distributed over the Moon's surface, so impact histories for the two sides are about the same. The absence of extensive spreads of lava in farside basins can scarcely be chance. It may be related to the tidal forces exerted by the Earth on a partly molten Moon early in its history. It may reflect the farside presence of a much thicker layer of crustal rocks through which there were no large-scale outpourings of lava like those through the comparatively thin-crusted nearside.

6 Farside cratered terra, from 116 km altitude

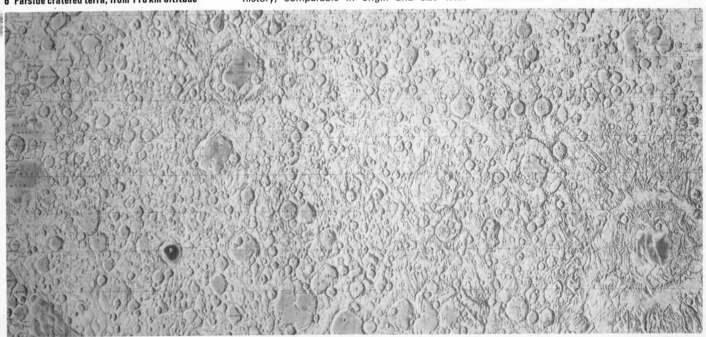

7 Central section of chart of Moon's reverse side, published 1970 at scale of 1 : 5 000 000

A closer look

Observations of the Moon's surface by powerful telescopes and by spacecraft photography reveal many different types of small-scale feature. There are craters of all shapes, sizes and stages of degradation. Some linear groups of small craters form *crater chains:* some fresh-looking craters have light-coloured surroundings, with *crater rays* (linear streaks) radiating outwards from their rims. Other linear features include *wrinkle ridges* (sinuous ridges) on the maria, *rilles* (canyon-like trenches) which cross the mare plains and floors of many large craters, and scarp-like *faults,* or surface dislocations, well displayed in landslip 'terraces' forming the walls of certain large craters. Small *domes* (circular-shaped protuberances), some capped by craters, may also be seen on the floors of large craters and in some mare regions.

A multitude of minor surface features, even individual rock boulders less than a metre across, can be distinguished on high-resolution photographs taken by spacecraft from lunar orbit. It is clear that many of these minor features are closely linked with the formation of craters. Most craters, large and small, together with most large boulders, boulder fields and many distinctive patterns seen in surface debris, are now attributed to the collision of meteorites with the lunar surface. They result from the explosive primary impacts of meteorites, or the secondary infall and redistribution of rock debris thrown up by the primary impacts. Some surface features are undoubtedly volcanic in origin, and have been interpreted as volcanoes, cinder cones and lava flows, but others, like the tracks of rolling boulders and 'patterned ground' (fig 42) observed on steep hillsides, are caused by downslope movements of rock debris under the influence of gravity.

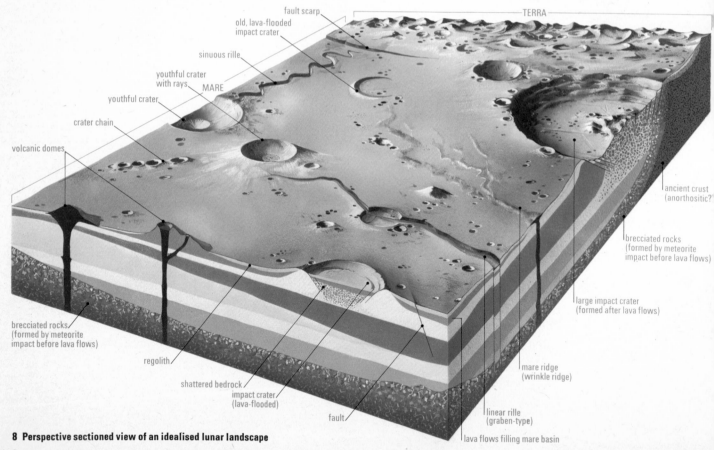

fault scarp

old, lava-flooded impact crater

sinuous rille

youthful crater with rays

youthful crater

crater chain

volcanic domes

brecciated rocks (formed by meteorite impact before lava flows)

regolith

shattered bedrock

impact crater (lava-flooded)

fault

TERRA

MARE

ancient crust (anorthositic?)

brecciated rocks (formed by meteorite impact before lava flows)

large impact crater (formed after lava flows)

mare ridge (wrinkle ridge)

linear rille (graben-type)

lava flows filling mare basin

8 Perspective sectioned view of an idealised lunar landscape

Craters of all sizes are the dominant feature of the Moon's surface. Even the deceptively smooth maria are pitted with small to very small craters or craterlets. Craters range from vaguely discernible shallow depressions (fig 45) to deeply excavated basins with sharply defined rims, some having terraced interior walls (figs 8, 10). Some are clearly old and degraded structures, partly obliterated or covered by later-formed craters and rock debris, whilst others are more youthful, fresh-looking features. Nearly all lunar craters were formed as a result of bombardment by cosmic debris either as *primary* impact craters, or as *secondary* craters produced by fall back of primary ejecta—the rock debris thrown up by primary impacts. Large primary impact craters, like Copernicus (figs 50, 51) and Langrenus (fig 10) often have flat floors, the central regions of which are occupied by high mountains, partly uplifted as a result of post-impact decompression. Volcanic craters are usually small and generally less regular in shape.

Crater chains are straight-line groupings of three or more closely spaced craters. The close linear relationship of craters within a single chain implies that they were all formed by a common process at the same time—either by impact excavation or by explosive volcanic activity. Most crater chains consist of comparatively small craters, excavated by the impact of rock debris ejected from larger impact craters elsewhere: each chain represents the surface infall of debris propelled in one direction, along slightly varying trajectories. Their secondary impact derivation is most clearly revealed by the radiating patterns they form around parent impact craters. Linear groups of small craters along some rilles (see p 8) are not the result of impact excavation, but are probably the surface expressions of volcanic activity along lines of crustal weakness or fracture. However, a prominent rille-like crater chain northeast of the primary impact crater Copernicus (figs 12, 51) is interpreted as a line of secondary impact craters associated with the formation of Copernicus.

Domes are small swellings on the Moon's surface produced by igneous activity. They vary in shape and size from simple blister-like, gently sloping hills—less than four kilometres across and only a few tens of metres in height—to large, less regular, steep-sided dome-shaped features, often showing cratered summits. These lunar domes, like their counterparts on Earth, were constructed in two different ways: *external growth* by successive eruptions of lava (forming small volcanic domes), and *internal growth* by sub-surface intrusion and build-up of molten rock, which, when cooled, gives rise to *laccoliths* (dome-shaped intrusions). More than four hundred domes have been detected on the lunar surface, most commonly in border regions of the maria (fig 11). They also occur on crater floors, especially in large impact craters like Copernicus and Tycho, but only rarely in highland terrain. The formations of domes spans a wide range of lunar history, being closely linked with the formation of mare basins and large impact craters (p 23).

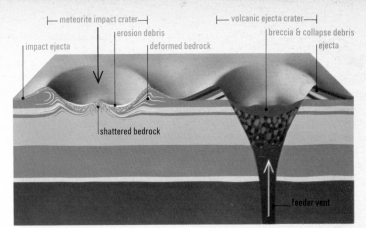

9 Origin of craters by meteorite impact and vulcanicity

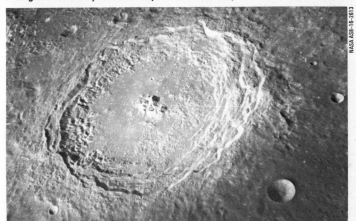

10 Langrenus, a 135 km-diameter impact crater in M. Fecunditatis

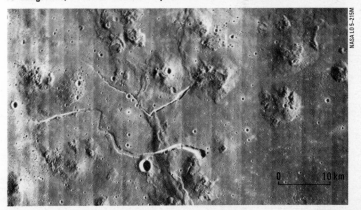

11 Orbiter 5 view of domefield in Oceanus Procellarum

Rilles, or *rimae*, are narrow trench-like features of the Moon's surface. They vary in width up to about five kilometres and, although rarely more than a few hundred metres deep, may extend as linear, arcuate or sinuous depressions for hundreds of kilometres. Straight rilles (fig 8) are thought to be 'graben'—narrow blocks of country dropped down between two parallel faults by tensional fracture of the lunar crust, particularly in response to the cooling of mare lavas. The more abundant sinuous rilles (figs 13, 14, 32) are downslope meandering channels which almost invariably start in small craters. Generally restricted to the maria they are regarded as ancient lava channels or collapsed lava tubes down which molten rock once flowed from volcanic vents.

Wrinkle ridges are sinuous elevations, up to about 30 km wide and rarely more than a few hundred metres in relief, some of which extend for hundreds of kilometres. Mostly observed on the maria (figs 4, 12, 15), the segmented form and dyke-like outcrop of typical mare ridges suggest that they originated in fissure eruptions, or in vulcanicity along faults. They may have formed as compressional upwarps in cooling lava flows associated with the last stages of filling the mare basins. The surfaces of broad mare ridges reveal small impact crater densities comparable with the adjacent maria—which confirms their contemporaneous formation. Ridge-like protrusions over highland terrain probably resulted from compression of semi-molten volcanic or impact-melted rocks.

13 Sinuous rilles near 40 km-diameter crater Aristarchus (centre left)

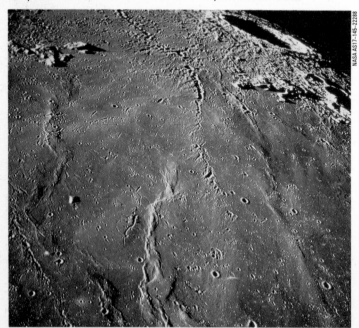

12 Ridges and crater chain on M.Imbrium, northeast of crater Copernicus

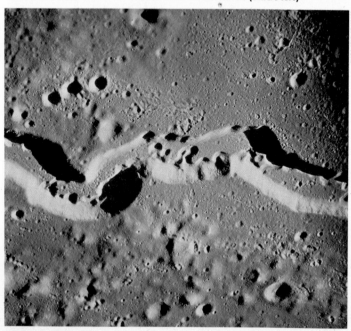

14 Detail of sinuous rille 8 km wide, near crater Aristarchus

Crater rays are long, brightly reflecting surface streaks which radiate from some craters, especially those of more youthful origin like Copernicus and Tycho. Such rays are extensions of equally bright 'apron' regions surrounding the craters (figs 2, 16, 17, 51). Both rays and aprons are areas of impacted rock debris ejected from parent craters and associated secondary impact craters. In general the brighter appearance of rayed areas when illuminated vertically by the Sun is due to large numbers of small secondary impact craters, many elongated parallel to ray trends. Another contributory factor is the presence of reflective glass particles (p 22 & fig 57) in the ejected debris: rock fragments vaporised by impact or vulcanicity pass through a liquid phase and chill as glass whilst explosively impelled.

Faults are dislocations in the lunar crust responsible for a variety of linear features. Graben-type subsidence or block faulting is considered to lead to the formation of certain types of rille (p 8). The inner walls of many impact craters (figs 10, 50) show concentric fractures and terrace-like fault blocks caused by extensive gravity slumping of wall materials. Faults with vertical ('dip-slip') movement and faults with horizontal ('strike-slip') movement are probably manifested in offset segments of crater walls, as discontinuities in mountain ranges and valleys, and as linear scarps. The Straight Wall (fig 15) is the scarp surface, 150 km long, of a normal dip-slip fault with a downthrow of about 250 metres to the west.

NASA AS13-60-8675

16 Farside rayed crater, Bruno Bright, from lunar orbit

PIC DU MIDI OBSERVATORY

15 The 'Straight Wall' fault escarpment in southeast M.Nubium

NASA AS15-97-13156

17 Rayed 'splash crater', from 118 km lunar orbit

Man on the Moon

The six Apollo landings, 1969 to 1972

18 **Historic confrontation of astronauts at Tranquillity Base, 21 July 1969**

Apollo 11

Men first set foot upon the Moon with twin goals of landing and returning safely to Earth, and securing a piece of the Moon's surface (the 'contingency sample') for later laboratory studies. In the event two Apollo astronauts accomplished a great deal more during four man-hours of activity on the lunar surface. They took many photographs; they collected over 21 kg of rock samples; they deployed scientific equipment around the landing site, including instruments to measure movements within the Moon (seismometer), to reflect laser beams back to Earth for accurate determination of the Earth–Moon distance (laser reflector), and to measure the type and energy of atomic particles from the Sun (solar wind panel). Some of the instruments continued to function long after the astronauts had left. The Mare Tranquillitatis landing site yielded rock samples which clearly established the fragmental nature of surface materials. Here, pulverised rock debris lay to a depth of five metres; the surface, though pock-marked by small craters up to 600 m in diameter, had altered little in 100 million years. Here were discovered fragments of basaltic lava which had cooled and crystallised more than 3700 million years ago.

20 Earthrise over lunar surface; orbiting LEM in foreground

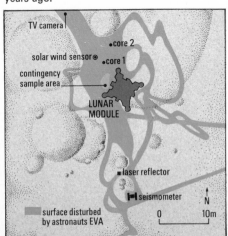

19 Traverse map of Apollo 11 landing site

21 Close-up camera on cratered regolith

22 Astronaut, seismometer, laser reflector and LEM

Apollo 12

The landing site selected on Mare Cognitum provided an environment similar to that experienced by the Apollo 11 crew, and enabled direct comparisons to be made between two mare surfaces in terms of materials, age and composition. Mission objectives included the more systematic geological exploration and sampling of the landing site (within a distance of 0.5 km from the Lunar Module), and the emplacement and activation of scientific equipment—the Apollo Lunar Surface Experiments Package or ALSEP. This included a seismometer, a magnetometer (for determining the magnetic field on the Moon's surface and the electrical conductivity of its interior), and instruments to gather information on the interaction of solar particles with the Moon's surface and almost negligible atmosphere. In their 15.5 man-hours of surface 'extra vehicular activity' (EVA) two astronauts collected over 34 kg of rock samples, including individually documented specimens, core-tube and trench samples. These surface materials, mostly basaltic igneous rocks and a few fragmental breccias, proved broadly comparable with those from the Apollo 11 site, although some basalt samples had originated in lava flows poured out almost 500 million years later.

24 Fully deployed ALSEP; magnetometer (foreground) and seismometer (right of astronaut)

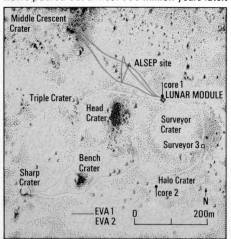

23 Traverse map of Apollo 12 landing site

Middle Crescent Crater

ALSEP site

Triple Crater

core 1
LUNAR MODULE

Head Crater

Surveyor Crater

Surveyor 3

Bench Crater

Halo Crater
core 2

Sharp Crater

N

EVA 1
EVA 2

0   200m

25 Astronaut using rock-sampling tongs

26 Impact-generated blocky clumps of regolith

27 Astronaut inspecting Surveyor 3 spacecraft, soft-landed 30 months earlier, with Lunar Module on horizon

28 Modularised Equipment Transporter and close-up camera (near elliptical shadow of aerial) on well-trodden regolith around Lunar Module

Apollo 14

Although lying only 180 km to the east of the previous Apollo landing the site selected for the Apollo 14 mission was in more rugged, heavily cratered terra bordering northeast Mare Cognitum. Pre-mission studies of photographs had suggested that the surface there might be composed of rock debris derived from the cataclysmic meteorite impact that had excavated the Imbrium basin, 1000 km in diameter. This conclusion was later supported by the astronauts' description of the landing area and by the nature of rock samples returned to Earth. Most of these samples proved to be fragmental, breccia-type rocks, consistent with an impact-generated origin, which had been formed about 4000 million years ago—much earlier than the basaltic rocks at the previously visited mare sites. During 18.3 man-hours of exploratory 'extra vehicular activity' on the Moon's surface two astronauts collected 43 kg of rock samples, deployed a range of scientific equipment (including an ALSEP experiments package, similar to that left by the Apollo 12 crew), and made a geological traverse 3 km long involving systematic descriptions, photography and rock sampling. A hand-drawn cart (MET) facilitated the transport of samples and tools.

30 Lunar Module over regolith crater

29 Central station for ALSEP (beyond crater)

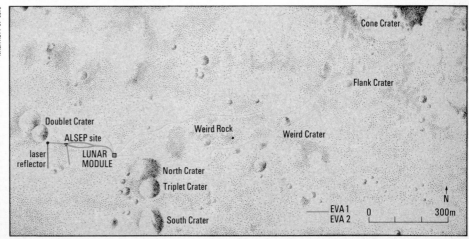

31 Traverse map of Apollo 14 landing site

Apollo 15

This mission, to the foothill region of the Apennine Mountains bordering southeastern Mare Imbrium, vastly increased the scale of lunar exploration. In addition to greatly extended surface observations, rock sampling and ALSEP experiments, a number of orbital investigations, including precision photography, X-ray and gamma-ray geochemical sensing, gathered data on the relief and chemical composition of the Moon's surface. The landing site lay between mountains that rose 4000 m above it—uplifted during impact formation of the Imbrium basin—and the valley-like Hadley Rille that cut into the surrounding terrain. During 37.1 man-hours of surface exploration, much helped by use of a battery-powered Lunar Roving Vehicle (LRV) the two astronauts travelled more than 28 km and collected 78 kg of rock samples. Most proved to be breccias, including and associated with fragments of basaltic lava and anorthosite (p 26), and revealed a complex history of impact events before the Imbrium cataclysm about 4000 million years ago. A core sample 2.5 m long of surface regolith showed nearly sixty different layers, each probably representing fine debris from a meteorite impact, which record the last 2500 million years of lunar history.

33 Farside crater Tsiolkovsky, with central mountains and mare-type floor

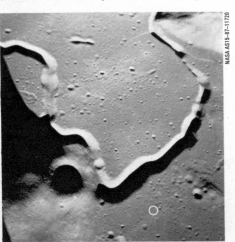

32 Hadley Rille and Apollo 15 landing site

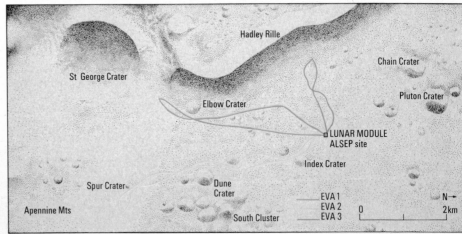

St George Crater

Hadley Rille

Chain Crater

Pluton Crater

Elbow Crater

LUNAR MODULE
ALSEP site

Index Crater

Spur Crater

Dune Crater

EVA 1
EVA 2
EVA 3

Apennine Mts

South Cluster

N

0   2km

34 Traverse map of Apollo 15 landing site

35 Astronaut and Lunar Roving Vehicle, seen against Mt Hadley rising 4 km above foreground plain

Apollo 16

Exploration of a small part of the rugged upland terrain lying to the west of Mare Nectaris, near the crater Descartes, was the prime objective of the Apollo 16 mission. Its complement of both orbital and surface experiments was almost identical to the previous mission: broad-scale geological, geochemical and geophysical mapping of the Moon's crust from orbit was continued, whilst detailed surface investigations in the landing site area were again considerably extended by emplacement and activation of Earth-monitored ALSEP experiments. During their 40.5 man-hours of surface exploration the two astronauts travelled more than 27 km over hummocky and cratered terrain, returning 96 kg of rock samples to the Lunar Module. Many proved to be fragmental breccias containing fragments of anorthosite. Studies of these samples showed a long history of meteoritic fragmentation and reworking of an ancient anorthositic crust before major basin-forming and cratering impacts ceased about 3900 million years ago. Thus after the Apollo 16 mission came a better understanding of the sequence of early basin formation and complexities of terra geology in the Moon's nearside southern highlands.

37 Cratered surface of M. Nubium; 115 km orbit

36 Panorama from Flag Crater: astronaut, LRV (upper left) and Stone Mt, 500 m high, on skyline (right)

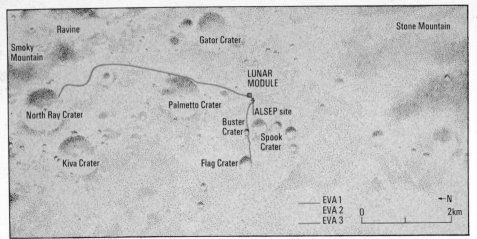

NASA AS16-118-18924

Ravine

Stone Mountain

Gator Crater

Smoky
Mountain

LUNAR
MODULE

Palmetto Crater

ALSEP site

North Ray Crater

Buster
Crater

Spook
Crater

Kiva Crater

Flag Crater

EVA 1
EVA 2
EVA 3

0   2km

←N

38 Traverse map of Apollo 16 landing site

39 Farside cratered terra; 116 km orbit

NASA AS16-114-18422, 23, 25, 27

40 Astronaut sampling large breccia boulder ('Split Rock'), Lunar Roving Vehicle in foreground (left)

Apollo 17

The first series of manned landings on the Moon ended with the Apollo 17 mission, which was the most successful. The pattern of investigation and experiment set by the two previous missions was continued: geological surveying and sampling in a pre-selected area, deploying and activating surface ALSEP experiments, and the completion from lunar orbit of scientific and photographic records of the Moon's crust. The landing site was in the Taurus-Littrow region of the mountainous highlands bordering south-eastern Mare Serenitatis. Here, during 44.2 man-hours of exploratory activity, two astronauts travelled 35 km and collected 120 kg of rock samples. These included basaltic lavas, gabbros, anorthosites and breccias and proved to be the most mixed collection returned by any Apollo mission. More clearly than earlier-collected samples they outlined the main stages of the Moon's 4600-million-year history: breccias with crystalline constituents derived by impact-fragmentation of the most ancient rocks, breccias formed by impact-excavation of the Serenitatis basin, basaltic lavas which later filled the Serenitatis basin, and the surface rock debris impact-generated locally during the last 3800 million years.

NASA AS17-147-22549

42 ALSEP equipment and LRV, with North Massif, 2000 m high, 3 km away

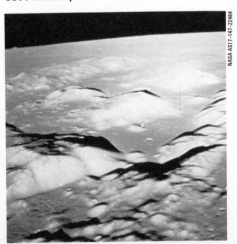

NASA AS17-147-22464

41 Landing site (centre) below orbiting Module

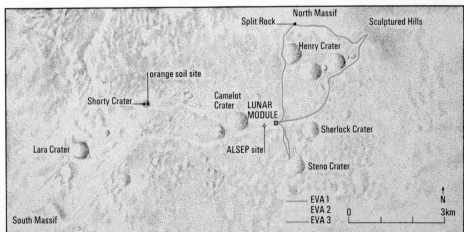

43 Traverse map of Apollo 17 landing site

Geology of the Moon

On the surface

The Moon's surface has been primarily shaped by impact cratering and volcanic activity, modified by a limited range of erosional processes operating in a vacuum-like environment (p 24). Almost everywhere the surface is formed by pulverised rock debris and fragmented meteorite material, mostly resulting from impact cratering over a long period of time. This superficial debris varies in nature and thickness from place to place, depending on local bedrock and history of impact activity. It may be several kilometres thick over ancient terra uplands, whilst on more youthful lava flows of the maria it is generally less than 50 m thick. The loosely consolidated upper layers of the surface debris form the lunar *regolith*. Rarely more than 20 m thick, the regolith mainly consists of various types of igneous rock, fragmental breccias, mineral grains and particles of glass (p 26). Although most of this material is fine-grained, weakly cohesive and soil-like, much of it less than 1 mm in particle size, the regolith also contains boulders, rocks and rock fragments of varying size, angularity and roundness. Local surface aggregations of blocks up to 15 m across are not uncommon. Everywhere the regolith surface is pockmarked by countless numbers of small impact craters, ranging from tiny pits to broad, shallow depressions in stages of degradation which vary with their age and the local impact history.

Stratification, or layering, of the regolith provides a record of this impact history. The sequence of layers, however variable or ill-defined, represents successive showers of rock debris, distributed and redistributed by impact events during many millions of years (p 16). Few outcrops of solid bedrock protrude through the regolith: none were encountered by the Apollo astronauts during their surface excursions.

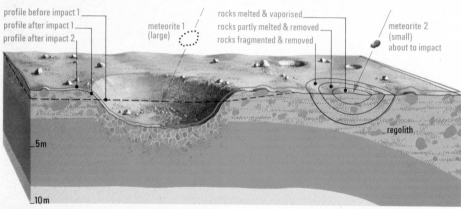

NASA AS11-40-5880

44 Cut-away perspective diagram of lunar regolith

46 Astronaut's bootprint in regolith, Apollo 11

NASA AS12-46-6734

NASA AS11-45-6704

45 Perspective view of mare surface at Apollo 12 landing site

47 Close-up of regolith, 72 × 83 mm

Vulcanism is the surface expression of igneous, or magmatic activity at depth. Molten rock or magma produced in the outer parts of the Moon (p 27) must have cooled and crystallised to form the primary igneous rocks of its crustal layers—some by simple gravity separation whilst molten, some by sub-surface igneous intrusion, and others by surface extrusion or vulcanicity. Many lunar surface features are now widely accepted as probably volcanic in origin. These include sinuous rille lava channels, pressure wrinkle ridges, domes, explosion vent craters and the lava flow infillings of large mare basins. Oblique solar illumination of some mare surfaces reveals what appear to be individual lava flows (fig 48). The floors of certain large craters like Copernicus and Tycho show dome-like mounds and a wide variety of patterns suggesting fluid-flow and shrinkage cracks in once-molten rock : small, isolated and pond-like flat areas may represent ancient lava lakes, or 'playas'. However, many of these crater features can also be explained by local production and movement of molten rock as a result of primary meteorite impact. Evidence currently available from isotope dating of lunar rock samples indicates widespread magmatic, igneous and volcanic activity between 4600 and 3800 million years ago, mostly before the excavation of the mare basins. A second major phase of activity between 3800 and 3200 million years ago led to the lava infilling of the mare basins. Most if not all of this regional vulcanism must have ceased before 3000 million years ago.

Impact cratering has been the most significant geological process in creating the Moon's surface and relief. An unceasing rain of meteorites has produced primary and secondary impact craters of all dimensions, ranging from huge basins 1000 km in diameter to tiny millimetre-size pits on individual rock fragments. In consequence, the igneous rocks of the Moon's crustal layers have been extensively shattered, modified and redistributed as fragmental surface rubble. Impact cratering not only gives rise to fragmentation of igneous bedrock but also causes shock melting and vaporisation of the rocks, together with other localised metamorphic effects, according to the scale of impact. Part of the vapour and melt forms globules of glass in flight ; some solidifies as glassy matrix which bonds fragmental rock debris into breccias. Both are impact-generated materials of widespread distribution in the lunar regolith (p 26).

The flux, or average rate, of meteorite bombardment of the Moon has declined with the passage of time. Most of the large craters and mare basins were impact-excavated during the first 1000 million years of lunar history (fig 28). Since the spread of volcanic lava flows over the floors of the mare basins ceased about 3000 million years ago relatively few large impacts have deeply cratered their surfaces. However, the mare regolith, with its small surface craters and finely-layered interior, reveals countless numbers of minor impacts during this period. Now the regolith surface everywhere must be in a state of virtual equilibrium : fresh impacts destroy on balance as many craters as the new ones they produce.

48 Fronts of lava flows picked out by low-angle sunlight, Mare Imbrium

49 Impact craters of different ages : large central crater 45 km in diameter

Weathering and erosion of the Moon's surface takes place in the total absence of atmosphere and water. Thus there is no terrestrial-type sub-aerial weathering by rain, wind, frost, and oxidising chemical reactions. There is no transport of erosion debris by rivers, glaciers, lacustrine or marine currents as occurs on Earth, and no accumulation of water-borne or wind-borne sediments to give rise to stratified rocks. Under the weak influence of lunar gravity (only a sixth of the Earth's gravity) there is very slow levelling out and redistribution of surface debris from higher ground—a process known as 'mass wasting'. This downward drift tends to fill craters, smooth out slopes and gradually subdue surface irregularities. No doubt the process is accelerated by impact and seismic vibrations, and by expansion–contraction movements of surface rocks as a result of the great temperature variation (250°C) between lunar day and night.

Protons and electrons streaming outwards from the Sun bombard the Moon's surface and may dislodge, or 'sputter', atoms from particles of lunar rock. As most lunar rocks consist of silicate minerals (p 26) these sputtered atoms will include oxygen, silicon and various metals. In general, the effects of 'solar wind' sputtering are the gradual rounding of surface blocks and fragments of rock—scarcely distinguishable from the 'sand-blasting' effects of micrometeorite impacts. A 10-cm rock could be pulverised by sputtering and small-particle impacts in about 50 million years: footprints (fig 46) may persist for only a few million years.

Stratigraphy deals with the relationships, order of superposition, and age of formation of recognisable rock groups and surface structures. A 'stratigraphical column' has been compiled for the Moon, following terrestrial principles that younger, later-formed rock groups and surface features generally lie above, obscure, and are more sharply defined than, older formations. On this basis the broad pattern of lunar history is reflected in four major, comparatively easily distinguishable stratigraphic divisions, or 'systems'—*Copernican* (the youngest), *Eratosthenian*, *Imbrian* and *Pre-Imbrian* (the oldest). These groups of rock-stratigraphic units, first defined by studies of astronomical telescope pictures of the Imbrium basin and its environs (fig 52), were later applied to all regions of the Moon. Recently, the Pre-Imbrian group has been subdivided into *Nectarian* and *Pre-Nectarian* systems, separated in time by the impact event which created the Nectaris basin about 4200 million years ago. Presumed Pre-Imbrian rocks and features appear in the mountainous terrae of the Moon's nearside and over much of its farside. Imbrian material includes features and rock debris resulting from impact-excavation of the Imbrium basin about 4000 million years ago: the Apennine and Carpathian Mountains were partly uplifted and mantled by ejecta at this time. Imbrian volcanic lavas occupy most of the large earlier-formed impact basins. Later impact craters and associated ejecta deposits have been grouped into Eratosthenian and Copernican systems, distinguished from each other by their field relationships and states of preservation.

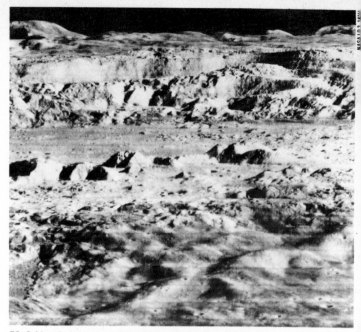

50 Orbiter 2 view across crater Copernicus, altitude 45 km

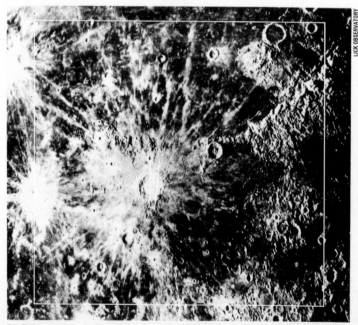

51 Telescope view of Copernicus region showing limits of figs 50 and 52

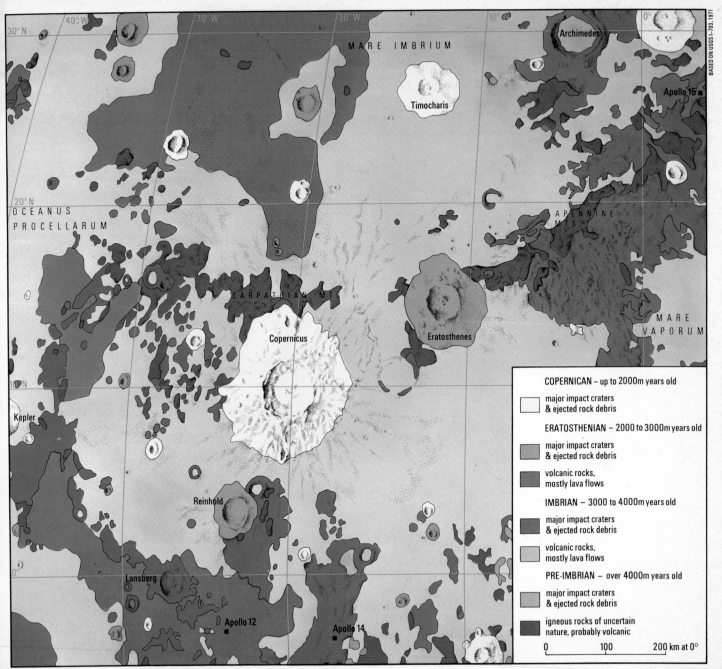

Map labels:
MARE IMBRIUM
Archimedes
Timocharis
Apollo 15
OCEANUS
PROCELLARUM
APENNINE MTS
20°N
Eratosthenes
MARE VAPORUM
CARPATHIAN MTS
Copernicus
Kepler
Reinhold
Lansberg
Apollo 12
Apollo 14

COPERNICAN – up to 2000m years old

major impact craters
& ejected rock debris

ERATOSTHENIAN – 2000 to 3000m years old

major impact craters
& ejected rock debris

volcanic rocks,
mostly lava flows

IMBRIAN – 3000 to 4000m years old

major impact craters
& ejected rock debris

volcanic rocks,
mostly lava flows

PRE-IMBRIAN – over 4000m years old

major impact craters
& ejected rock debris

igneous rocks of uncertain
nature, probably volcanic

0   100   200 km at 0°

52 Simplified geological map of the Copernicus region

The lunar rocks

Igneous rocks, formed by cooling and crystallisation of molten silicate material, have been of paramount importance in shaping the surface geology of the Moon. They are present either as solid primary rocks, or as the constituents of secondary rocks and regolith (p 22). Allowing for minor but significant chemical differences lunar igneous rocks are broadly comparable with varieties known on Earth, and include fine-grained basaltic lavas (grain size less than 1 mm), coarser-grained diorites, gabbros and feldspar-rich anorthosites (grain size greater than 1 mm). Three distinct groups have emerged from analyses and age determinations carried out on lunar samples returned to Earth. The first group comprises the anorthosite–norite–troctolite (or 'ANT') gabbroic rocks and high-alumina basalts of the terrae, all of which crystallised from 4600 to 4000 million years ago. The second group comprises basaltic rocks rich in potassium (K), containing unusually high amounts of rare-earth elements (REE) and phosphorus (P). These are the so-called 'KREEP' basaltic rocks which solidified from 4000 to 3800 million years ago. The third group comprises mare basalts, or 'FETI' basalts, comparatively rich in iron (Fe) and titanium (Ti) mostly extruded as lava flows from 3800 to 3200 million years ago.

Blanketing many areas, and widely dispersed in the regolith, are many varieties of impact-generated breccia (fig 56). These fragmental rocks range from lumps of glass-bonded regolith debris to the more extensive and thicker deposits of consolidated debris formed by large, basin-excavating impacts (fig 58). In these latter, the so-called 'ejecta blanket breccias', the shock-melted rock material has slowly cooled and annealed to form a partly crystalline bonding matrix. The loosely compacted and soil-like regolith contains a wide range of rock and mineral fragments together with many forms of once molten glassy material (figs 57 bd).

53 Lunar basalt and photomicrograph

55 Lunar anorthosite and photomicrograph

54 Lunar gabbro and photomicrograph

56 Lunar breccia and photomicrograph

26

What the rocks reveal about the Moon and its past is now the subject of worldwide streams of research publications. Never before in the history of geoscience have so few rock samples been subjected to analysis and measurement of almost every conceivable chemical and physical property, by specialists utilising all the resources of modern technology. Many questions have been answered by these researches, but a host of others have been posed. *We now know* that solid rocks first crystallised on the Moon at least 4600 million years ago—that the lunar rocks and their minerals (feldspars, pyroxenes, olivine, silica phases, opaque oxides and others) crystallised and consolidated in the complete absence of water—that there are no living or fossil organisms in lunar rocks—that there are significant chemical differences between comparable igneous rocks of Earth and Moon. These discoveries, among many others, apply fresh limitations on future conjecture and hypothesis in almost all branches of lunar science.

57 Lunar rocks (a) vesicular basalt (b) soil constituents (c) glazed breccia (d) glassy 'marbles' in soil

(A) NASA 15016 (B) ZEISS (C) NASA 15255 (D) NASA S69-45728

The Moon's interior has been shown by seismic ('moonquake') and other geophysical evidence to be zoned into a rigid crust, a thick solid mantle and a partly molten core (fig 58). The nearside crust varies in thickness from 30 to 45 km under the maria and from 30 to 70 km under the terrae, increasing to almost 100 km on the farside. Large high-density mass concentrations, or 'mascons', have been detected beneath the surface of some circular mare basins. With an average density of 2.7 g/cm³ the feldspathic rocks of the terrae are lighter than the KREEP rocks (p 26), and are much lighter than the mare basalts which approximate 3.3 g/cm³—the overall density of the Moon. These crustal rocks must represent the products of chemical and gravity separation, or 'differentiation', from once partly-molten mantle material of an olivine-pyroxene composition. The core might be chemically similar to the lower mantle, though alternative models which satisfy the geophysical data feature iron, or iron-sulphur cores of different radii.

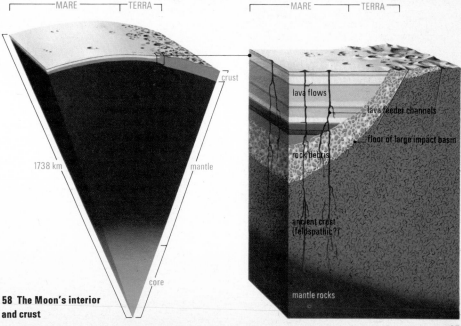

58 The Moon's interior and crust

The Moon's life story

MOON   EARTH

meteorite impacts on large scale

widespread volcanic activity gives
rise to atmosphere and oceans

oldest known rocks

first life in the oceans

meteorite impacts and vulcanicity
continue on much reduced scale

PRE-IMBRIAN ② ③
4000 million years ago

IMBRIAN ④ ⑤
3000 million years ago

PRECAMBRIAN

surface topography now changes
little with passing time

ERATOSTHENIAN
2000 million years ago

① origin by accretion

② first crust formed

③ huge mare basins formed by impact
of large, asteroid-size meteorites

④ volcanic activity, mare basins filled by lava flows

⑤ youngest known volcanic rocks

meteorite impacts on small scale
only few large meteorites

COPERNICAN

1000 million years ago

The Moon may now represent an early stage in the history of the continuously
evolving Earth, all earthly traces of which are now obliterated. The Moon's
development as a planet may have been arrested at this stage because of
inadequacies of size and internal heat: a comparatively cool and rigid interior
never gave its outer layer the mobility shown by the Earth's crust: a low gravity
never allowed the retention of an atmosphere which would have partly
protected the lunar surface from the constant infall of meteorites.

crater Copernicus formed

immobility of crust, coupled with absence of atmosphere
and almost negligible erosion, ensures little change in
surface features and rocks over aeons of time.

How was the moon formed?

Earth and Moon formed as planet and satellite by separate
accretion, over 4600 million years ago, condensing from vast
cloud of cosmic matter with other members of the Solar System.

Moon was once part of a primordial Earth. It was thrown out
into space from the Earth, which was rotating more rapidly than
at present, and gradually moved out to orbit now occupied.

Moon was formed as independent planet, at the same time as
other members of Solar System. The planetary orbit passed
sufficiently close to Earth for Moon to be captured as a satellite.

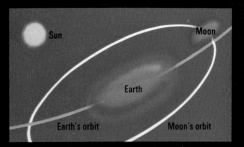

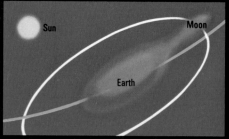

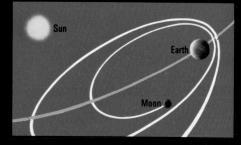

Compared with the Earth

the Moon has a cold, thick and rigid crust, unlike the Earth's comparatively thin, mobile crust *so there has been no folding or large-scale metamorphism of lunar rocks*

the Moon has about half the heat flow through its crust *so, being much smaller, must have more heat-generating radioactive elements in its outer layers*

the Moon has a weak magnetic field *so does not have a large nickel-iron core like the Earth*

the Moon has a low level of moonquake activity (or natural seismicity) *so has no internal convection movements such as are inferred in the Earth's mantle*

the Moon has no surface water *so there are no water-lain sedimentary rocks and no erosional processes involving water*

the Moon has no life *for the development of life on Earth was mainly due to the presence of surface water and atmosphere*

the Moon has no appreciable atmosphere *so has negligible atmospheric oxygen and almost no oxidation of rocks*

the Moon has similar igneous rocks but with significant chemical differences *so it probably had a separate origin and history*

the Moon has no igneous rocks so far dated at less than 3000 million years old *so cannot have many (if any) geologically youthful features of internal origin. Igneous rocks on the Earth are mostly less than 3000 million years old*

the Moon has igneous rocks dated as 4600 million years old *but the oldest rocks on the Earth so far dated are 3800 million years old – and these are very rare*

continuous renewal of oceanic (basaltic) crust causes opening and closing of ocean basins, with 'drift' and reshaping of continents: mountain building; vulcanicity; earthquakes

increase in atmospheric oxygen due to photosynthesis by marine organisms

CAMBRIAN to RECENT

animal life evolving rapidly in oxygenated environment

single large continent of Pangaea begins break-up leading to present-day geography

crater Tycho formed

present day

Moon and Earth to correct scale

The planet Mars

Mean distance from Sun
226.56 million kilometres

Period of rotation around Sun
1.88 Earth years

Period of axial rotation
1.03 Earth days

Diameter
6775 km mean

Mass
0.108 of Earth's mass

Density
3.89 in terms of water
0.70 of Earth's mean density

ARCADIA

AMAZONIS

Olympus Mons T H A R S I S

MESOGAEA

North Spot

Middle Spot

ZEPHYRIA

South Spot

Valles Marineris

MARE SIRENUM

Solis Lacus

MARE
ERYTHRAEUM

MARE
CIMMERIUM

ARGYRE

The moons of Mars Phobos,
(above), the larger of the two
Martian satellites, has a maximum
diameter of 22 km and orbits at a
mean distance of 9270 km.
Deimos (right), has a maximum
diameter of about 12 km and
orbits at a mean distance of
23 400 km.

Mariner 9 American orbiting
spacecraft which transmitted
back to Earth over 7300 pictures
of the Martian surface between
November 1971 and October 1972.

Viking Lander American spacecraft
successfully soft-landed on Mars in 1976
for direct investigations on its surface

59 The planet Mars and its moons

As viewed by spacecraft

Mars reveals a surface shaped by meteorite impacts and by volcanic, erosional and depositional processes. In some regions it is cratered like the Moon, whilst elsewhere it has gigantic volcanoes and steep-walled canyons, much larger than any on Earth. It is a world etched by wind erosion, patterned by wind-borne sediments, channelled by fluid erosion no longer evident, and topped by seasonally variable polar 'icecaps'.

Craters of impact origin are widespread over Mars, ranging from colossal basins like Argyre and Hellas – 1800 km in diameter and 4 km deep (fig 65) – down to the smallest hollow. Most of the craters appear to be eroded and old, with low rims, shallow interiors and no ejecta blankets or rays (fig 60). Cratered terrain dominates the southern hemisphere and constitutes almost half the equatorial region. The sparsely cratered northern hemisphere reflects less a different impact history than substantial modification by volcanic and erosional processes.

Volcanoes are amongst the most spectacular features of the Martian landscape. In the Tharsis region of the northern hemisphere are four huge 'shield' volcanoes, the largest of which, Olympus Mons (fig 62a), is a pile of lavas 25 km high and 600 km in diameter, with a caldera-type summit crater 65 km in diameter. Small, cratered volcanic domes have also been recognised, mostly in the Tharsis region (fig 62b). The southern hemisphere has fewer, smaller and more eroded (older?) volcanoes.

Canyons and tributary ravines of vast dimensions form an east-west belt of 'canyonlands', Valles Marineris, about 2500 km long, near the Martian equator. The main canyon is generally between 100 and 200 km wide, and plunges to depths exceeding 5 km in places. Part of it (fig 61) is almost 400 km across and displays deeply gullied walls with mounds of landslip debris at their base.

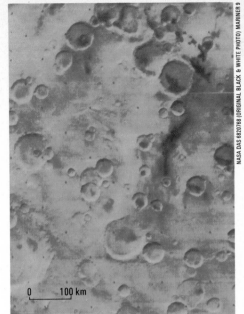

60 Old cratered terrain

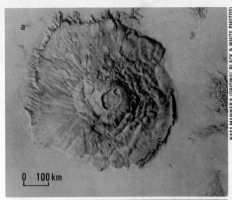

62 Volcanoes: (a) Olympus Mons (b) Tharsis 'dome'

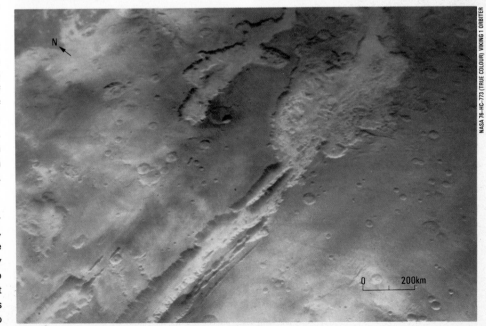

61 Canyonlands: part of Valles Marineris seen from 30 000 km

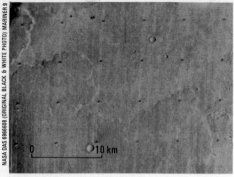

63 Volcanic terrain: lava flows

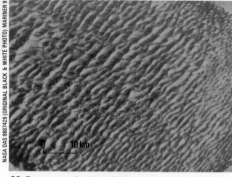

66 Desert terrain: dunefield

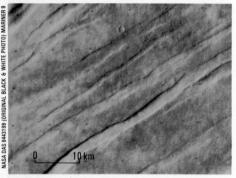

64 Fractured terrain: graben and horsts

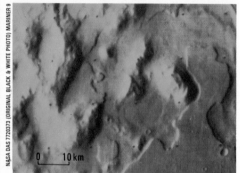

67 Etched terrain

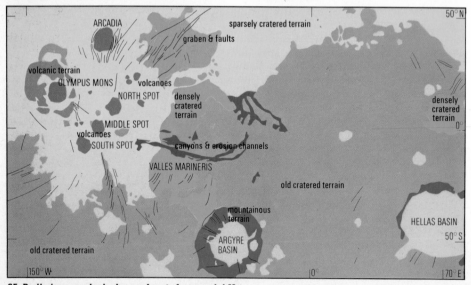

65 Preliminary geological map of part of equatorial Mars

Faults and rifts are varieties of tectonic fracture in the Martian crust. Many sub-parallel and radially aligned faults—some up to several thousand kilometres long—are associated with the shield volcanoes and domes, especially in the Tharsis region and around Arcadia (fig 65). Some of these faults let down linear rift-like blocks of country ('graben') between upstanding blocks or 'horsts' (fig 64). Such crustal fracturing is probably a tensional adjustment to vertical pressures which result from localised volcanic uplift.

Deserts of one kind or another—rocky, sandy dusty, or of polar 'ice' (mostly water ice with some frozen carbon dioxide)—cover Mars, although the term 'desert' was formerly applied to the brighter reflecting regions between the dark 'maria' (fig 59). However, all the Martian landscape is a cold windswept desert, with wind-sculpted rock features partly obliterated beneath loose rock fragments, windblown sand and dust (fig 70). In places dunefields have been formed much as on Earth, but with dunes more closely spaced near field margins (fig 67).

Martian geology has been tentatively assessed and mapped, using a classification of surface features or 'terrain units' based on physical and structural characteristics deduced from spacecraft photographs. These terrain units can be arranged in a stratigraphic sequence, which allows a tentative interpretation of the planet's geological history. Some terms used in naming the terrain units, like 'volcanic' and 'densely cratered', have geological connotations. Others, like 'etched' and 'laminated', are descriptive and do not specify a geological process or the materials involved. However, the hollows and pits of etched terrain (fig 67) probably result from intense wind erosion of older terrains. Laminated terrain (fig 69), bordering the polar regions, appears to consist of layered sediments over 5 km thick. These could be wind-eroded beds of volcanic ash or morainic dust left by major retreats of the icecaps (fig 68).

Compared with Earth and Moon

Our present knowledge of Mars leads to the view that the planet may represent a transition in its evolution and geology, from a relatively primitive, impact-dominated, dead world like the Moon, towards a more mature, partly water-covered and atmosphere-enshrouded planet, volcanically active and with a more mobile crust, such as the Earth.

Mars, like the Earth and Moon, probably endured intense bombardment by meteorites during its early history over 4000 million years ago: the well-cratered Martian terrains, like the lunar highlands, probably survive from this period. But unlike the Moon—dead for the past 3000 million years—the formation of shield volcanoes and lava plains has probably spanned most of Martian history, much as on Earth. However, the vulcanism which gave rise to the huge shield volcanoes on Mars seems to indicate a lack of the crustal mobility which characterises the Earth : the piles of lava are so vast on Mars because they accumulated in a few places over long periods of time. This means that Earth-type crustal plates and plate movements (leading to folding and metamorphism of rocks) have not yet developed on Mars, although rifting and faulting on Mars appear to be Earth-like. The Marineris canyon system (fig 61) may have originated as a rift-type graben like the smaller East African rift system on Earth.

The thin Martian atmosphere consists of carbon dioxide (95%) with some nitrogen (2–3%) and traces of water vapour, argon and oxygen. Although tenuous by Earth standards (a surface pressure of 7 millibars compares with about 1013 millibars on Earth) the Martian atmosphere allows some solar energy to be converted into wind activity. Hence, unlike the Moon, Mars has terrestrial-style wind erosion, oxidation and weathering of surface rocks, transport and deposition of fine rock debris, and formation of stratified sedimentary rocks. The erosional activity of water on Mars at some time in its history is suggested by sinuous surface channels comparable with water-eroded features on Earth. Indeed, large quantities of water may still be present in the form of 'permafrost' (frozen groundwater) beneath the Martian surface.

As seen from Viking Lander 1 (fig 70) the Martian surface is similar to many rocky, partly sand-covered terrestrial deserts. Chemical analyses of the surface debris suggest the presence of iron-rich basaltic lavas comparable with varieties known on Earth. The red colour of rocks and soil is probably due to highly oxidised iron-bearing minerals. No life has been detected.

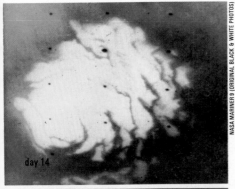

day 14

day 94

68 South polar 'icecap' summer retreat

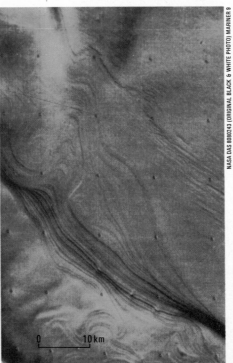

0   10 km

69 Laminated terrain

70 Viking Lander view of the Martian landscape

Meteorites
Rocks from the depths of space

71 **Artist's impression of meteorite fall and fireball**

Meteorites are rock fragments from inter-planetary space which chance has brought to Earth. They are of three types: *stones*—by far the most abundant (92.8% of observed falls), *irons* (5.7%), and *stony-irons* (1.5%). Stones largely consist of silicate minerals—like olivine, pyroxene and feldspar—and other minerals known in terrestrial and lunar igneous rocks. Over 85% of stones are 'chondrites' (fig 72), distinguished from other igneous rocks by the presence of small spherical inclusions of silicate material, called *chondrules*. Iron meteorites are essentially alloys of iron with up to 20% of nickel (fig 75). Most irons consist of two nickel-iron minerals in lamellar intergrowths which show up as criss-cross patterns on etched and polished surfaces (fig 76). Stony-iron meteorites are composed of nickel-iron and silicates in approximately equal proportions: some show the nickel-iron enclosing discrete grains of olivine (fig 73). Many stones and stony-irons show smooth and furrowed crusts (fig 72) as a result of *ablation* (superficial melting) as they passed through Earth's atmosphere. Some irons show scalloped depressions so formed (fig 75).

Isotope dating of meteorites reveals mineralogical ages of about 4600 million years, as old as the oldest dated Moon rocks (p 29), and equal to the calculated age of the Earth and presumably the other planets. They probably represent primordial planetary materials, the aggregation and differentiation of which ceased about 4600 million years ago. However, the great diversity of meteorites precludes their origin from a single differentiated parent body. Most meteorites probably formed much later when small, but different-sized parent bodies in the asteroid belt between Mars and Jupiter collided and broke up.

Tektites are enigmatic fragments of glassy material found in certain regions of the Earth's surface (figs 74, 78). Although widely interpreted as meteorites, showing features of atmospheric ablation, they may be fusion products of cometary or meteorite impacts on Earth.

72 Stony meteorite, Barwell, England

75 Iron meteorite, Canyon Diablo, USA

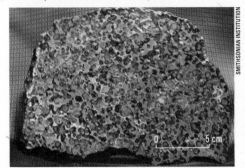

73 Stony-iron meteorite, Springwater, USA

76 Polished & etched iron, Coopertown, USA

Georgia tektite
Czech moldavite
Java tektite
Philippinite
Texas bediasites
Indochinites
Australites

74 Some well-known tektite varieties and forms

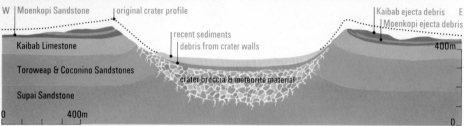

W | Moenkopi Sandstone | original crater profile | Kaibab ejecta debris | E
Moenkopi ejecta debris

Kaibab Limestone

recent sediments
debris from crater walls

400m

Toroweap & Coconino Sandstones

crater breccia & meteorite material

Supai Sandstone

0   400m   0

77 Meteor Crater, Arizona, USA: view from south and cross section W–E

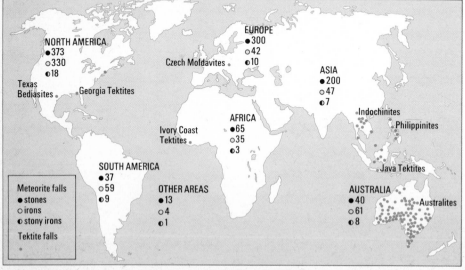

EUROPE
● 300
○ 42
◑ 10

NORTH AMERICA
● 373
○ 330
◑ 18

Czech Moldavites

ASIA
● 200
○ 47
◑ 7

Texas
Bediasites   Georgia Tektites

Indochinites

Philippinites

AFRICA
● 65
○ 35
◑ 3

Ivory Coast
Tektites

Java Tektites

SOUTH AMERICA
● 37
○ 59
◑ 9

OTHER AREAS
● 13
○ 4
◑ 1

AUSTRALIA
● 40
○ 61
◑ 8

Australites

Meteorite falls
● stones
○ irons
◑ stony irons

Tektite falls

78 World distribution of known meteorite and tektite falls up to 1962

Striking the Earth. Thousands of meteoroids enter the Earth's atmosphere every day. Most are small, almost dust-like particles and are rapidly incinerated by the intense frictional heat of high-speed atmospheric flight. Their incandescent deaths, marked by brilliant streaks of light, are the 'shooting stars' or 'meteors' of the night sky. Only few of the larger meteoroids or their fragmented remains survive the violent passage through the atmosphere to reach the Earth's surface as meteorites, and even then about two-thirds of them fall into the oceans. Rarely more than ten meteorites are recorded and recovered annually as falls. A meteorite fall may be observed as a bright fireball with long, incandescent trails of ablation debris (fig 71). It may be heard as a thundering, whistling or cracking sound, sometimes accompanied by loud 'supersonic shock-wave' bangs. Sites of meteorite falls, of which almost 2000 are now known, show a random distribution, but tektites appear to be confined to 'strewnfields' in certain geographically limited regions (fig 78): tektite falls have never been observed. Scars of meteorite impacts on the Earth's surface are rare, largely because normal geological processes gradually lead to their disappearance: only the larger craters survive any length of time, generally under favourable conditions of climate and surface geology, as exemplified by the 20 000-year-old Meteor Crater in Arizona (fig 77). However, systematic investigations in many parts of the world now suggest the presence of up to 60 structures of meteoritic origin, some of which are associated with fragments of nickel-iron and minerals showing the effect of high pressures consistent with impact modification.

Suggested further reading. *Geology of the Moon,* by T. A. Mutch, Princeton University Press, 1972; *Planetary Geology,* by N. M. Short, Prentice-Hall Inc., 1975; *Meteorites and the Origin of Planets,* by J. A. Wood, McGraw Hill, 1968. First results of the Viking missions to Mars are summarised in *Science,* Vol. 193, part 4255, and Vol. 194, parts 4260, 4271, 1976.